Charles Henry Land

The Scientific Adaptation of Artificial Dentures

Charles Henry Land

The Scientific Adaptation of Artificial Dentures

ISBN/EAN: 9783337420093

Printed in Europe, USA, Canada, Australia, Japan

Cover: Foto ©berggeist007 / pixelio.de

More available books at **www.hansebooks.com**

THE

SCIENTIFIC ADAPTATION

OF

ARTIFICIAL DENTURES.

BY

C. H. LAND.

DETROIT, MICH.:

PUBLISHED BY C. H. LAND.

1885.

PREFACE.

To treat systematically of the conditions found to exist in the mouth, and to properly adapt artificial dentures, is the object of this little book and to those who will give it their careful consideration it is respectfully dedicated.

■

CONTENTS.

INTRODUCTION.

WHAT is an adaptation? The most perfect contact to be produced in a mechanical way may be demonstrated with two pieces of polished glass. The surfaces are uniformly horizontal and present equally balanced points of contact representing the maximum results to be obtained in a strictly manipulative manner. In this particular there are two rigid substances, requiring but one rule to produce definite results. However by converting the horizontal surfaces into one of many angles, especially if they are rigid, the difficulty of securing perfect contact is increased tenfold, necessitating an entirely different mode of procedure. Thus we are compelled to resort to the use of plastic materials, and in this way can accomplish measurements, that the eye fails to see, from which to secure a denture that must represent an equally balanced contact on perpendicular, angular and horizontal surfaces. Here the profession have rested, believing that to obtain a good impression was about the limit of possibilites, but how different do we find it. The plaster impression is uniformly rigid, from which a rigid cast is secured, the denture the same; the conditions of the mouth are not indicated. The average practitioner sends the rigid impression to the laboratory, expecting no more than a rigid cast, a rigid denture and a rigid fit, satisfied that he has done his best. No careful diagnosis of the conditions of the mouth have been made, consequently no proper relief has been provided. If careful investigation and exact measurements will not accomplish more, then my efforts will be in vain. This I cannot believe. Therefore, in presenting these few pages to the profession, I shall offer no apology, believing that whatever will have a tendency to advance the cause of Dentistry will be thoroughly appreciated.

Respectfully,

C. H. LAND.

HORIZONTAL CONTACT.

WHEN two polished surfaces are brought together, their adhesion has been estimated to equal a retaining force of about 21 grains to the square inch. But when a fluid substance is placed between, the contact is so much more perfect as to increase it over one pound. The ordinary leather sucker will demonstrate a lifting force of about two pounds, this only on horizontal surfaces; on all angles it slips or slides, the intermediate fluid acting similar to a lubricator. This will account for the difficulty of fitting angular mouths. A positive distinction must be made between the angular, perpendicular and horizontal surfaces. The latter are the first to come in direct contact with the denture consequently offer the first permanent resistance to the denture. The dense portions of the maxillary bone and central parts of the palatine surface offer the first positive contact to the denture, and, unless some provision is made to prevent this uneven contact, the adaptation will be imperfect. The remedy is to relieve all horizontal surfaces except the outer fifth. If rigid, the relief should be slight. A soft condition of the alveolar ridge demands a much greater central relief; this will permit the denture to come in more direct contact with the greatest diameter of the mouth, and secure a balanced contact. A mouth that is uniformly rigid is so nearly allied to a rigid impression, a rigid cast, and a rigid denture, as to represent a balanced contact demanding but the slightest modification.

ANGULAR SURFACES.

WHEN the horizontal surfaces are relieved the angular surfaces are brought in more positive contact, thus producing a balanced adaptation. A mouth that is deep and narrow presents the minimum horizontal surface and the maximum angular, consequently indicates the most feeble retaining force. The intermediate fluid causes the denture to slip or slide on the angles. To remedy this difficulty in addition to relieving the horizontal surfaces, we must resort to friction; this can be done by trimming the cast on all angles that are perpendicular. In this way a lateral pressure is created and is equal to one count in favor of a better adaptation. By this means the mouth will be forced into a smaller space than the impression represented.

BALANCED CONTACT.

A FLAT mouth having no angles uniformly rigid will correspond to a uniformly rigid denture, and, in such cases, a perfect impression will nearly always represent a balanced contact, and needs but the slightest modification in order to produce a perfect fit. An angular mouth, uniformly rigid, and a rigid denture would not be well balanced; an immediate resistance would be found on all horizontal surfaces, which, if relieved slightly, would equalize the pressure and produce a more positive contact on the angles. Add to this lateral friction on all perpendicular surfaces, and the maximum results are obtained. In order to secure a balanced contact in the average mouth, four-fifths of the interior surface of the denture must be relieved, the direct pressure will then come on the alveolar ridge. By establishing the first positive contact on the outer fifth, the inner four fifths will be the last to conform. The mouth gradually yields to this central relief until the entire surface is equally balanced. Positive contact on the central fifth means a denture that rocks or rides on the palatine surface. Positive contact on the outer fifth means a denture that cannot rock or ride on the center.

RETAINING FORCE.

THE amount of force or exertion manifested in the retention of artificial dentures varies according to the particular shape and conditions of the mouth. The amount of weight each will support ranges from ounces to pounds. A small, dry, and rigid mouth would not support more than from three to six ounces, on the contrary, the same mouth if moist and comparatively soft, would support from one to three pounds.

PHILOSOPHY OF THE ADHESION OF THE DENTURE.

WHEN two plane surfaces are brought together and an intermediate fluid placed between, they will adhere with a force equal to about two pounds to the square inch; this adhesion is due to the attraction that all fluid substances have for solid bodies. A drop of water will adhere to a pane of glass; the same attractive force will cause water to rise against its own gravity between two pieces of glass. In the same manner water will travel up a tube; hence the term capillary attraction. It is a manifestation of the same law that takes place between the denture and mouth. If placed in a vacuum the denture will adhere independent of the air pressure, proving that the retention is not due to the influence of the atmosphere. (See Adhesion, Appleton's American Cyclopædia, page 115.)

ATMOSPHERIC DENTURES.

THERE can be no manifestation of atmospheric pressure without an air space and some exhausting means. With an air pump and suitable appliances an atmospheric pressure of 14 pounds to the square inch can be demonstrated. Take an ordinary student lamp chimney, cover a two-pound weight with rubber dam, press the chimney close to the weight, then with the mouth exhaust the air, and it is with the utmost difficulty the weight can be lifted. The large end of the chimney is 1½ inches in diameter; therefore, admitting that one pound pressure to the square inch was secured, how long would the tender tissues of the mouth tolerate it? Has not experience shown that eventually all air spaces become filled and at this juncture atmospheric pressure ceases to exist? Again, when any fluid substance enters the air spaces the possibility of a vacuum is destroyed. Both the fluid and tissues of the mouth will invariably fill all air spaces, thus annihilating even the chance of a partial vacuum.

When any substance of a more dense nature fills the cavity, it takes the place the air formerly occupied. In this way an equilibrium is established, and offers a counter-resistance to the air pressure in the same manner as capillary vessels in our body relieve us of the weight of the atmosphere. From these facts it will be seen that the term Atmospheric Dentures is a misnomer. All dentures are practically maintained by the adhesion of an intermediate fluid. Upon it depends the permanency of the adaptation. At best the utility of the atmosphere is but transitory. As a temporary means to establish an adaptation there is some value in it, estimating the exhausting power of the tongue to be about 6 ounces to the square inch, and having relieved four-fifths of the interior surface of the denture (yet this cannot be called an adaptation until the exhausting power of the tongue has forced the mouth to conform to it), close contact will then be equally distributed over the entire surface, and not until this space is filled will perfect adaptation be secured. From this it will be seen that, if we had to depend on the atmosphere to maintain artificial dentures, it would be a complete failure. It is fortunate that the fluids of the mouth come to our rescue to save our patients from the inevitable injury that atmospheric pressure would produce. The tissues of the mouth are too sensitive and tender to tolerate a partial vacuum, even if it could be provided.

FOUR-FIFTHS RELIEF.

IN forming reliefs in artificial dentures four-fifths is made the standard, from the fact that, by relieving almost the entire central portion of the mouth first, contact is thrown to the outside and maximum adaptation is secured. Any rigid portion of the mouth coming in direct contact with a rigid denture will destroy a perfect adaptation. These parts are usually found in the central portion of the palatine surface and the dense parts of the maxillary bone, . and, unless some relief is given, the denture will rock or ride either forward, backward or sideways. By reference to the following illustration it will be seen that L L C C C mark the position of a lead or tin matrix to form a proper relief that will be effective in all cases where the alveolar ridge is soft and spongy and the central portion

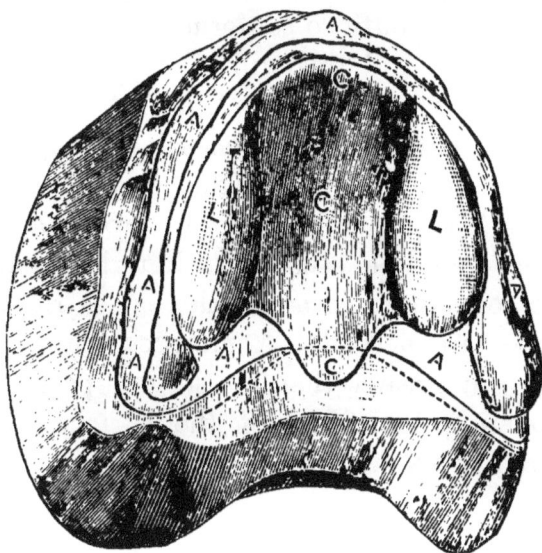

2

of the palatine surface rigid as far back as the soft palate. The lobes L L are designed to give an additional relief to the dense parts of the maxillary bone, and are about the thirty-second part of an inch in thickness, with edges beveled so as to avoid any acute angles. The central portion, C C C, is about the sixtieth part of an inch thick, and, when greater depth is needed, an additional piece of lead or tin is inserted beneath this portion. The dotted lines indicate the extent of the denture, showing that, when completed, there will be a space or opening at C. This serves as a complete relief and permits the air to pass between the denture and the mouth, in such cases and in order to get an immediate adhesion, it will be necessary to insert a moist piece of cotton batting at this particular point until the mouth so conforms to the denture as to close the space, and this will take place in from three to four days. The letters A A A A A indicate the parts of the mouth that, when soft and spongy, will bear increased pressure, and this is brought about by trimming the plaster cast just in proportion to the conditions found in each case; if rigid, no trimming will be tolerated. In a mouth that is rigid and flat no lobes are needed. A matrix of uniform thickness, say not greater than the sixtieth part of an inch and covering four-fifths of the surface, edges beveled, would give the best results.

ADAPTATION OF LOWER DENTURES.

IN the proper adjustment of a lower denture the points of contact are quite the reverse of the upper. In fitting a denture to the superior maxillary arch, the aim is to relieve the central portion so that the positive contact is thrown to the outside; on the contrary, with the lower it is absolutely necessary to relieve the outside and have the positive contact directly on the center of the alveolar ridge, this being the only portion that will bear increased pressure, especially where the ridge is thin and sharp. The tender membrane at the sides will not tolerate even the slightest friction of the denture, and it is common to see dentists cutting away the denture until it is quite as narrow as the hard portions of the ridge. This is a great mistake; the best plan is to bevel the edge even to the depth of ¼ inch, this will relieve the soft tissues and not make it necessary to so narrow the denture that it becomes useless. It is essential to keep the denture just as deep as possible on the side toward the tongue, the surface should be slightly concave, this will allow the flexible portion of that member to have more room, and at the same time prevent the too ready movement of the denture from side to side. Also, the food will not be so apt to get beneath it. On the outside the same principle will hold good, the denture should be made concave on the surface and be just as wide as the cheek will permit it to be without any undue friction, this, also, will have a tendency to hold the denture more firmly in position by the soft muscles falling into the concave surface, a considerable capillary adhesion is secured thereby, producing results far superior to cutting away and narrowing the denture. When the denture is made wide enough so that it comes in slight contact with the cheek, the soft tissues will be expanded, and the fluids of the mouth (obeying the laws of adhesive attraction) will cause the tissues to adhere with a force sufficient to hold the denture firmly in position.

ADAPTING PARTIAL DENTURES.

IN the adjustment of partial dentures, owing to the limited horizontal surface, the adaptation by capillary contact will be feeble, and in forming a relief it should cover only the horizontal portion and be very slight. Also, by trimming the cast at the side a lateral pressure will be of some advantage, especially on perpendicular surfaces. Close adjustment to the remaining natural teeth is important, especially on the bicuspids and molars, but not on the incisors, the latter should be slightly relieved on the lingual surface, as they are most always horizontal and come in contact with the denture.

THE UTILITY OF THE MUSCLES OF THE CHEEKS IN ASSISTING TO MAINTAIN A DENTURE.

A LL previous discussions on adapting dentures have been strictly confined to the alveolar arch and ridge, the soft tissues and muscles of the cheek have not been considered as of any value, but on the contrary, in many respects, have proved to be detrimental. However, a careful study of the action of these flexible and soft tissues will demonstrate the possibility of making them remarkably useful in helping to support both the upper and lower denture. It is a common error to allow the denture to fit too close on the tender portions of the alveolar ridge, just where the soft tissues of the cheek unite with the denser parts; and, still more unreasonable, it is generally reduced to a knife-like edge which, in many cases, terribly lacerates the mouth. The idea seems to prevail that the denture must be as deep as possible. A better way is to have it just as deep as these soft and flexible tissues will tolerate, perfectly free from friction, on the horizontal edge, which, in place of being sharp, should be nicely rounded and thoroughly polished. Just over the bicuspids make the edge not thinner than the 32nd part of an inch, and the least in depth. Then, over the molars it should

be from ¼ to ⅜ of an inch or more, according to the
mouth. Also just over the cuspid it should be ¼ of
an inch thick. Then the surface over the molars and
cuspids should be made concave, the cheek will be
drawn into the concave surface, and, being expanded
over the thick edge, consequently increase the retaining power of the denture about fifty per cent. The
same results can be secured on the lower jaw by con-
caving the sides of the denture, the cheeks will fold
into them and, in many cases, create a decided capillary adhesion where least expected.

DIAGNOSIS OF THE MOUTH.

THE importance of a thorough examination of the
mouth previous to constructing a denture can
not be over-estimated, and yet, by the average practitioner, very little attention is shown it. Therefore as
an aid, to those who will take sufficient interest, the
following illustration will serve as a guide by which
the various conditions may be so systematized as to
become of the utmost value. Secure a plaster cast of
the mouth, then, with pencil in hand, note the conditions in a similar manner as shown in the illustrations.
The letters R R R indicate that within the outlines it
is rigid; A A A, that the tissues are not quite so
rigid; C C C C C C, tissues soft and flexible; B B B,

very tender. Pay particular attention to the part
marked 32. 16. 32, see if the tissues, when an instrument is pressed on the parts, will yield to about the
16th and 32nd part of an inch in depth.

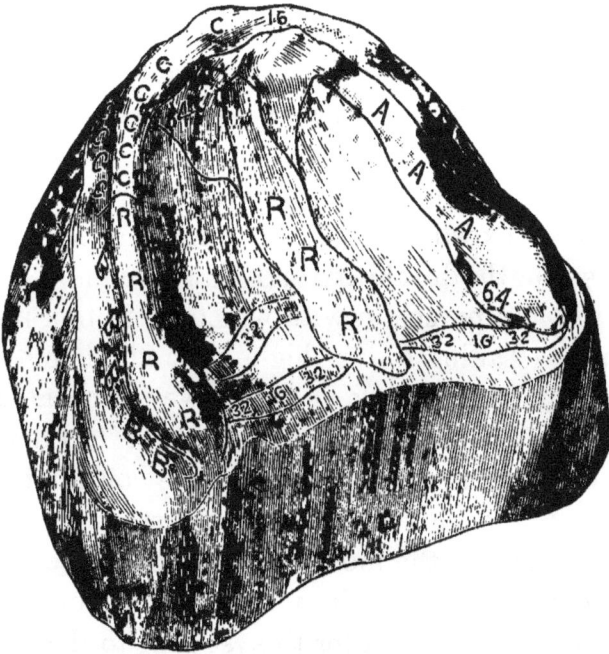

To the casual observer the merits of this system
will not be appreciated until a little experience will
teach the value of the training it will give, enabling
one to become thoroughly familiar with many peculiar
features that were not noticed before, and be the only
means by which perfect results can be obtained.

SELECTION OF MATERIAL ACCORDING TO THE CONDITION OF THE MOUTH.

O UT of various materials that have been used to form a base for artificial dentures, the practical field of selection may be reduced to four: porcelain, rubber, gold and platinum; these have proved to be the standard of excellence. Rubber for cheapness; gold for its strength, durability and unlimited application; platinum and porcelain with which perfection in art is obtained. Taking for my highest standard continuous gum and porcelain work, and the lowest rubber. In the selection of materials for the mouth for full dentures, continuous gum and porcelain is to be considered first, gold next, and rubber last. The first estimate to be made is, whether the mouth will support a set of continuous gum or porcelain, if not, would combination gold and rubber do, after this submit to the rubber? The rules to guide may be summed up as follows:

A mouth that is small, rigid and flat, or rigid and angular, would not represent a sufficient retaining force to support a continuous gum set, the selection therefore should be rubber and plain teeth, and whenever the lips are not long enough to hide the appearance, gum sections can be used. In adapting a denture to a mouth that represents the minimum retaining force, to use the lightest material is one count in their favor; to make a four-fifths relief is two counts; to utilize the muscles of the cheeks is three, and a proper articulation is four. If full advantage of these very important factors is thoroughly utilized, a mouth with a feeble retaining force can be made entirely satisfactory, when it would be a complete failure if any *one* of these factors were left out.

Finally the rule is to recommend continuous gum work, whenever the conditions of the mouth are most favorable for a strong retaining force, and rubber when they are the least favorable. A rigid denture and a rigid mouth demand rubber; 1st, it is light and strong; 2nd, it will have a tendency to soften the tissues, the very condition necessary for a better adaptation. When the mouth is soft there is a rebound to it, so that when a little extra pressure is thrown to one side the opposite will not let go so quickly. With a rigid mouth and a rigid denture, there being no rebound to the tissues, the instant the denture is pressed on one side it will let go on the opposite.

THE ADAPTATION OF PORCELAIN WORK.

THE value of a system of manipulating the plaster cast by which there will be no necessity of grinding will be recognized at once. This can be

done by becoming familiar with the following facts:
Let it be understood that a horizontal piece of porcelain would shrink just one-fifth, and, it will be seen,
that on a horizontal mouth the adaptation would not
be interfered with, unless on the outer margin there
was some perpendicular sides that would be brought
closer together in consequence of the central shrinkage. Notice all angles that are perpendicular especially; those that are anterior to the alveolar ridge
would be drawn one-fifth closer to the center.
Therefore in manipulating the cast add one-fifth
additional plaster on all perpendicular surfaces anterior to the alveolar ridge, this will allow for the
shrinkage toward the center. But on the lingual
surface, especially where inclined to be perpendicular,
the cast must be trimmed away one-fifth, more par-
ticularly on the posterior portion of the lingual surface of the alveolar ridge. By reference to the following illustration the num-
bers 1–16, 1–32, indicate
that in addition to trimming
the posterior parts a groove
corresponding in depth can
be cut in the cast, here the
tissues are soft and flexible
and will bear increased pres-
sure. This groove will form a rib that is intended
to press into the soft parts where they are inclined to
perpendicular, but in no case on the horizontal sur-

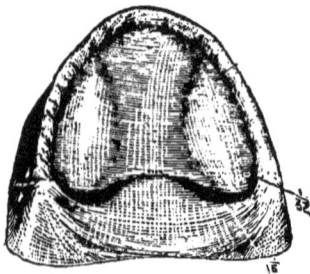

face; this rib is formed at the posterior portion of the central relief as shown in the engraving. Porcelain, like all other bases, has its place in Dentistry, and can not be used indiscriminately. A wide and rigid mouth, with a lower set of natural teeth to antagonize, would be almost certain to fracture the denture, the same with continuous gum. In such cases combination gold and rubber attachment would be the best. Where the lower antagonism is artificial, the danger is not so great, and entire sets of porcelain have proved to be a very satisfactory piece of work, and, with a better acquaintance of its excellent qualities together with the modern appliances for manipulating it, there seems to be no reason why it may not become the most popular base.

[From the *Independent Practitioner*.]

A CRITICAL REVIEW ON ATMOS-PHERIC PRESSURE, ETC.

By C. H. LAND.

THE entire force of the argument of Dr. —— is concentrated on what he asserts is a powerful pressure of the atmosphere. This, it sems to me, is an absurdity. For how can there be any manifestations of atmospheric pressure without an air space and some exhausting means? These ideas seem to pervade almost the entire dental profession, and, as a consequence, if one goes into any dental laboratory, not excepting those of the colleges, he will find the ubiquitous air chamber staring at him as an emblem of stupidity. It may be seen on signs, on cards, bill heads, letter heads, and in the newspapers. The beautiful little heart, shield, or star, is the vehicle of last resort for the suction-plate dentist, who, for the want of a better training in the proper method of inserting a relief, places a leaden heart in the most rigid portion of the palatine surface, for no other reason than because he saw his perceptor do it. In order to enable others to test the extent of this pressure, let me suggest the following:

1st. Take a glass chimney, such as is used for the ordinary student's lamp, get a two-pound weight, cover it with rubber dam, press the large end of the chimney on the weight, and with the mouth exhaust the air, and it will be demonstrated that the power exerted by the tongue as an exhausting means, under the most favorable conditions, will not represent a retaining force greater than about one pound to the square inch.

2d. Take a piece of leather one and one-half inches in diameter, fasten a string in the center, soak it in water, then press it close to a six-pound weight, and a retaining force of two pounds' adhesion to the square inch will be manifested, a contact force one hundred per cent. greater than where the tongue is employed as an exhausting means. In this latter experiment we have a manifestation of capillary attraction, induced by the adhesive attraction that all fluid substances have for solid bodies, to such an extent that the water will act against its own gravity, as illustrated in a sponge. It will force the air out of the intermediate spaces and take its place. In this way an equilibrium is established, by which the atmosphere meets with a counter-resistance sufficient to completely balance the weight, consequently there is no pressure. The saliva, or fluids of the mouth, cause the denture to adhere with a retaining force of about two pounds to the square inch, forcing the air out from beneath the denture, and the tissues of the mouth fill up all the empty spaces. From these facts it should become evident to the most casual observer, that the term atmospheric denture is a misnomer. However, if we could utilize the pressure of the atmosphere through the agency of an exhausting pump, even with one-pound pressure to the square inch, there could be no better means devised for maintaining a denture. There would be this practical advantage: the air pressing equally on all sides would cause the denture to be held firmly in position from all directions, no matter what the shape of the mouth, while a capillary contact will slip or slide on the angles, and only represent its maximum retaining force on horizontal surfaces. For this reason angular mouths are the most difficult to fit, especially if they are rigid and dry.

The scientific facts that I wish to present to the profession may be enumerated as follows:

1st. That artificial dentures are not maintained by the pressure of the atmosphere, and, if they were, that the tissues of the mouth would not tolerate even one pound pressure to the square inch.

2d. That they are maintained through the agency of an intermediate fluid, which represents a retaining force of about two pounds to the square inch on horizontal surfaces.

3d. That air spaces should be provided, not for the purpose of forming a partial vacuum, but simply as a relief to the dense portions of the maxillary bone and the rigid parts of the palatine arch. This relief, in a mouth where the alveolar ridge is comparatively soft, should cover four-fifths of the interior surface of the mouth, and be just sufficient in depth to make the outer fifth become the most positive in contact. The mouth being of a yielding nature, it soon conforms to this central relief, and, as a result, we secure a balanced denture, having established the first positive contact at the greatest diameter of the mouth. A mouth that is generally rigid, representing a surface so closely allied to a rigid denture, would indicate a balanced contact. Consequently, a good impression is about all that is necessary to secure perfect results.

On the contrary, a mouth that presents certain rigid parts in combination with soft or flexible tissues is an indication of the necessity of relieving the denser portions on all horizontal surfaces, in this way producing a balanced contact. When properly manipulated this should reward the dentist with the utmost possibilities of success in the application of a rigid denture to the dental arch.

HINTS FOR THE LABORATORY.

Assays, Weight, and Value of Gold and Silver Coins.—The sterling values given omit the value of the alloy used, which is never considered in practice.

GOLD COINS.

Country.	Name of Coin.	Fineness.		Standard Weight of Coin.	Weight of pure Gold.	Net Value Sterling.	
		Carat.	*Grains*	*Grains.*	*Grains.*	*s.*	*d.*
England...	Sovereign..	22	0	123¼	113	20	0
France....	Napoleon..	21	2¼	97½	89½	15	10
U. S......	Eagle......	21	3½	268½	246	43	6½
India.....	Mohur....	22	0	180	165	29	2½
Austria....	Ducat.....	23	2¾	58⅙	53⅛	9	5¼
Prussia.	Frederick, 1800.	21	2	100½	92¼	16	4

SILVER COINS.

Country.	Name of Coin.	Fineness.		Standard Weight.	Pure Silver.	Net Value.	
		In 12 parts.		*Grains.*	*Grains.*	*s.*	*d.*
England...	Shilling...	11	2	93	86	0	$11\frac{6}{10}$
France....	Franc.....	10	15	75	69½	0	$9\frac{4}{10}$
U. S......	Dollar.....	10	13½	400	370	4	2
India.....	Rupee.....	11	0	178½	165	1	10¼
Austria.	Rix dollar., 1800.	9	17	384	355½	4	0
Prussia.	Convention Rix dollar..	9	19	388	359	4	0½

To reduce sovereigns to lower stands, add for each coin—

For 16 carat + 46 grains alloy.
"   18   "   " 27½   "   "
"   20   "   " 12   "   "

For other coins, take the weight of pure metal in the table, calculate the proportion of alloy required, and deduct from this the amount of alloy the coin already contains, as shown in the table.

Acids for Cleaning Plates and Scraps.—For gold and plantinum, two parts hydrochloric acid to one of water. The commercial acid is as good as the pure for all purposes, if free from arsenic. For dental alloy, equal parts sulphuric acid and water. Sulphuric acid should be free from lead. For cleaning filings, boil in nitric acid until dry, wash and melt, throwing in a small quantity of salpetre or corrosive sublimate before pouring. Acids for cleaning plates should be kept in enameled cups, so that they can be readily heated if necessary.

Anvil.—When the workshop is not on the ground-floor, the anvil should be placed on india-rubber blocks about one inch thick, to stop vibration and noise, and the face of the anvil should be surrounded by a raised ring, to enable side blows to be given with certainty.

Blowpipe Support.—The best material is a mixture of equal parts of finely ground and sifted charcoal with an equal weight of china clay, made into a mass with rice-flour paste, moulded to shape and dried. It is clean, a good non-conductor, and everlasting. These can now be purchased at the depôts.

Box for Gold-plate, Teeth, &c.—The most convenient form is a box say 10 in. square with a handle on the top, opening at the side only, with a door. This should be filled with shallow wood or tin trays sliding on guides. The tray for scraps, &c., should have several tin boxes fitting in it, in which scraps, filings, band gold, &c., can be kept separate.

Casting.—A very small quantity of chloride of calcium added to the water with which the sand is moistened will keep it permanently damp, and save trouble. This need only be done once.

Zinc.—If this gets thick, make it red-hot, and then pour about half an ounce of strong hydrochloric acid on it, stirring at the same time with an iron poker. A dross will separate in a few seconds, and the zinc will be quite fluid, and fit for use.

Zinc is considerably improved for casting purposes by the addition of one-fourth of its weight of grain tin.

Cement Sticks for holding bands and teeth in position.—A mixture of dried Canada balsam and bees-wax. If the Canada balsam is soft, it should be heated until it sets hard on cooling.

Cement (thick) for fastening pivot-teeth, repairing models, &c.—A solution of gum copal in methylated ether. This will be found one of the most valuable cements for general use, and should always be kept ready for use in both labratory and operating room.

Cement (thin) for general purposes, and varnishing models.—A solution of gum mastic in methylated spirit.

Cement, mastic (so-called).—A solution of gum sandarac in spirit. The latter gum is sometimes substituted for mastic by retailers from its cheapness. The readiest way to distinguish one from the other is to put a small piece in the mouth; sandarac remains brittle; mastic becomes tough and workable between the teeth.

Cement for surfaces which can be heated.—A small piece of mastic or shell-lac heated between the surfaces to be joined, which are pressed together when hot, make a firm and water-tight joint.

Cement for rendering boxes, &c., tight against water and weak acids (for boxes containing batteries, &c.)—A solution of sealing wax in methylated spirit.

Crucibles of Plumbago are, as a rule, damp when new, and require drying before heating up suddenly the first time; but they may afterwards be exposed to the most sudden changes without risk.

For gold melting these are the best and last much longer than fireclay. When foul with flux, they may be scraped off whilst red-hot, the flux being placed in acid to recover any beads of gold it may contain.

Disinfectant.—A very weak solution of permanganate of potash will destroy instantly any taint from diseased roots or imperfectly-cleaned plates, and should always be used to rinse the spitoon in hot weather every time it is made use of. It is cheap, satisfactory, almost tasteless, not poisonous, and quite free from smell. It may be satisfactory to some to know that this will remove the taint of smoking from the breath if used as as a mouth-wash.

Gold (fine) sp. gr. 19·3 (22 carat, sp. gr. 17.3)— The hardest copper alloy contains 7 gold to 1 copper. Copper may be dissolved out from the surface of gold-plate by a hot solution of ammonia, and tarnished plate may be cleaned and improved in color by the same means.

The malleability of gold is less affected by silver than by any other metal.

For ordinary gold-plate, coin gold and coin silver may be used without the addition of copper, of which they contain sufficient. For bands and wire, old pins, solder, backings, and clasps should be melted together, and if making new metal for clasps, the alloy should be equal parts coin silver and fine copper.

Glass may be filed about as readily as brass, by using a file kept moistened with spirits of turpentine, and may be drilled by the same means. A five-sided broach broken at the end makes the best drill.

Gravers.—Old oval mouth-files softened, filed to shape, and tempered to dark yellow approaching purple, make first-rate gravers.

Hands, to keep clean and soft.—Use twenty or thirty drops of an equal. mixture of strong liquid ammonia and glycerine in the water in which the hands are washed. It is of the utmost importance, if the hands are to be kept in good condition, that they should be carefully washed the last thing at night. This will do more than any amount of care expended on them during the day.

Hard Solder for Repairing Instruments.—An alloy of coin silver and copper in equal parts makes the best hard solder known. It works fully as well as gold solder, and is extremely tough.

Left Hand.—A few weeks steady training of the left hand, making use of it on all possible occasions, will enable it to be used for most purposes as freely as the right. The value and convenience of this power will well repay the trouble.

Lead.—The best for dental purposes is obtained by melting up the linings of old tea-chests. If required specially soft, it should be used hot, with a few folds of paper between the plate and reverse, to enable them to be separated easily.

Loam.—A mixture of plaster and silver sand answers well, and being free from iron, does not stain the teeth when heated, as ordinary sand is liable to do. Casting sand should not be used, as it contains clay, which warps and cracks when heated. A stiff copper frame helps to prevent warping. Silver sand may be obtained from seedsmen, who use it for raising cuttings of plants, or from glass-works, at about the same price as plaster.

Lubricating Oil.—Most of the highly praised oils contain spirit of turpentine or mineral oils. These, although apparently improving an ordinary oil, give it also the property of slowly hardening like linseed oil, and should therefore be most carefully avoided for the Morrison engine, automatic mallets, &c. The best oil is made by exposing fine sperm oil to bright sunshine along with a quantity of clean lead shav-ings. In two or three weeks the oil becomes almost as clear and fluid as water, and is the most perfect lubricant for lathes and delicate instruments. One drop in any ordinary bearing is ample; more, simply runs about and makes a mess.

Melting-points of Metals.—Lead, 612°; tin, 442°; zinc, 680°; bismuth, 476°; antimony, 810°; red heat, 980°; heat of common fire, 1140°; brass, 1870°; silver, 1873°; copper, 1996°; gold, 2016°; cast-iron, 2786°; platinum, 3280°. The temperatures are given for pure metals, without alloys.

Mercury, sp. gr. 13.5.—Commercial mercury may be purified by standing under a small quantity of very dilute nitric acid for two or three weeks, or by boiling in the same for two or three hours. It may also be rendered pure by agitating in moderately strong sulphuric acid until the acid no longer be-comes turbid, or takes up any foreign substance.

The amalgam of silver and mercury expands slowly for some days after it has become hard.

Nickel-plating.—A smooth deposit of nickel may be obtained by using a moderately strong solution of chloride of nickel and ammonia, with moderate bat-tery power, and nickel or platinum anodes.

Nickel-plating without Battery.—In a porcelain or copper vessel place a concentrated solution of zinc chloride diluted with 1½ times its bulk of water, and boil.   If any precipitate falls, it must be re-dissolved by the addition of a few drops of hydrochloric acid. A small quantity of powdered zinc must be thrown in, which covers the vessel inside with a coating of zinc.   Add sulphate or chloride of nickel until the liquid is distinctly green ; then put in the articles to be plated (after cleaning them thoroughly), along with a few fragments of zinc, and continue the boiling for fifteen minutes.   If a thicker coating is desired, the operation may be repeated.

Oil Bottles. — The best arrangement for oiling lathes is one of the drop-bottles, such as are supplied with oxychloride fillings.   They should be sunk in a recess in the lathe-top, or in a convenient position, and remain in the same place, so that they can always be found in a moment.

Plaster of Paris.—If spoilt, or badly burnt, use a solution of potash, either caustic, or the carbonate, or sulphate, instead of water.   Plaster which has set will set again with the above solution if ground up.

Plaster, to Harden.—Mix with limewater; this is not suited for vulcanite work.   Mix with solution of alum ; this is suitable for all purposes.

If a model already made is required harder, boil it for ten minutes in strong solution of alum.

Models of Irregularities or Rare Cases.—If these are required for specimens, use best plaster, mixed with a very strong solution of borax, dry the model carefully, and soak in melted paraffin which has been slightly tinged with gamboge and dragon's blood. If carefully done, the resulting model has the appearance of fine Italian marble, not having the slightest resemblance to plaster. These models can be cleaned and washed without injury. The borax is necessary to give the peculiar semi-transparent appearance.

Pins for Models.—Use straight, smooth, tinned wire, so that if a model is difficult to cast, the teeth may be cut off with a piercing-saw, and the tooth and pin withdrawn. This is a convenience when teeth are very irregular.

Platinum, sp. gr. 21, melting-point, 3280°.—Found on the western slopes of the Cordillera of the Andez, Brazil, Hayti, and several parts of Russia. Welds at white heat, and makes a perfect joint with glass at a red heat. Soluble in nitro-muriatic acid only. Alloys with most metals, some with great violence, and evolution of intense heat (tih, &c.).

Dental alloy may be made by adding finely-cut platinum scrap to melted coin silver, and exposing it to a white heat in Dr. Land's gas furnace for from fifteen to twenty-five minutes. Let the crucible cool, turn it over, and give it a sharp tap on a board; this will remove the button without breaking the crucible. Place the button wrong side up in the crucible, so that the platinum which may have settled at the bottom will be uppermost, and melt again. Before pouring, all platinum alloys should be thoroughly stirred with a clay-pipe stem or plumbago rod, or the resulting ingot will be irregular and not work well. An excess of scrap should be used, and what remains at the botom of the crucible used for the next melting. The longer the alloy is kept melted and stirred, the larger will be the proportion of platinum taken up. Coin silver, from its containing copper, gives a much harder alloy than fine silver, but an equal proportion of each may be used. The temperature required for melting good quality alloy is about the same as cast-iron, which is obtained in Dr. Land's gas furnace in ten minutes, and fireclay crucibles are preferable, as platinum combines with the carbon (carbide of platinum) of plumbago crucibles at a white heat, and for the same reason borax should be sparingly used, if at all. It is possible that the mixture of the brittle boride or carbide of platinum in ordinary dental alloy makes it so irregular and uncertain in its tensile strength. The silver can be dissolved out of the surface by hot oil of vitriol.

Alloys of Gold and Platinum.—1 platinum and 1 fine gold very malleable, and nearly the same color as gold. 1 platinum and 9.6 gold has the color of gold and the density of platinum. 1 platinum to 11 gold has the color of platinum.

Palladium, sp. gr. 11.5.—Welds at a white heat, infusible at a white heat, and similar in most properties to platinum, but at present far more costly, and therefore not used. The material recently sold as palladium contains only a small percentage of that metal, and is therefore inferior to common dental alloy. 1 palladium with 4 fine gold is white, hard, and ductile.

Precious Metals.—The Hall-marks on manufactured articles show where they have been assayed, and are as follows:—Birmingham, an anchor; Chester, three wheatsheaves, or a dagger; Dublin, a harp, or figure of Hibernia; Edinburgh, a thistle, or castle and lion; Exeter, a castle with two wings; Glasgow, a tree, and a salmon with a ring in its mouth; London, a leopard's head; Newcastle-on-Tyne, three castles; Sheffield, a crown; York, five lions and a cross.

The standard mark for gold of 22 carats and sterling silver is—for England, a lion passant; Edinburgh, a thistle; Glasgow, a lion rampant; Ireland, a harp crowned. Gold of 18 carats, a crown, and the figures 18; silver of the new standard (seldom used), 11 oz. 10 dwts. fine, a figure of Britannia. Duty mark (indicates that the duty has been paid) the head of the sovereign.

Rubber.—The more foreign matter this contains the less liable it is to warp after vulcanising. Ash's IX pink adheres strongly to clean gold-plate; red and black part readily from it, and will often warp and lift clear from a plate unless strongly fastened. If rubber will not vulcanise in contact with dental alloy, paint the metal with a strong solution of bichloride of plantinum, and make it red-hot. This will form a surface of metallic plantinum, and prevent the combination of the sulphur in the rubber with the silver in the alloy. A solution of chloride of gold will answer the same purpose, but is much more costly.

Riveting Hammers.—These should have the faces ground to a hemispherical shape, like a boiler riveter's hammer. A hemispherical head is easy and safe to use, and makes a rivet-head the same shape as itself.

Steel.—The best steel for fine instruments is Stub's, which can be obtained at almost all the tool-shops. Old files softened, filed or forged to shape, hardened and tempered, make first-rate cutting-tools, the steel being almost invariably of very fine quality. A good steel instrument ought to cut softened steel readily, and retain a good edge without bur or chip.

Sharpening Instruments.—Arkansas stone is the best, but difficult to get in good pieces. Washita stone is almost as good, and easy to get in large slabs at a moderate price. Do not mount them in a wood case, but use one side for water and the other for oil. For a very fine edge, use a strip of buff leather fastened on a stick, and charged with crocus and oil. Once charged, the leather requires no more for two or three years.

Stool for Work-bench.—If made with a revolving top, this will be found far more comfortable than the ordinary pattern. This can be done at little expense, by fastening the top to an upright iron rod working in a socket near the floor, and a collar near the seat. The few shillings expended on this convenience is soon recovered in the shape of clothes, which are rapidly worn by the continual movements necessary.

Spirits of Wine.—The methylated spirit sold generally contains a small quantity of gum or resin, which makes it troublesome for spirit-lamps, and, as a rule, is not sold free from resin, unless specially asked for. If the contents of any vessel are boiling over, a few drops of spirit sprinkled over the surface will stop the rising instantly. The strength of commercial alcohol may be increased by placing in a bladder, and hanging up in a current of air. The water contained will evaporate through the pores of the bladder, until absolute alcohol remains. Bladder is impervious to spirit, although it allows a ready passage to water.

Tempering Instruments.—Very delicate instruments may be tempered with certainty in a bath of melted lead, with an ordinary chemical thermometer graduated to 600°; pale straw, 430° F.; dark yellow, 470°; brown yellow, 500°, yellowish purple, 520°; light purple, 530°; dark purple, 550°; dark blue, 570°. Polished instruments may be tempered by laying on a block of hot iron until the color appears. Small excavators should be made red-hot, and stuck repeatedly into a stick of sealing-wax until cold. They will not require tempering afterwards.

Vulcanite, to Repair.—Roughen the surface thoroughly with a rasp, graver, or piercing-saw, and paint the surface over with a solution of Ash's pink rubber in chloroform. When this is dry, new rubber can be built on the surface and vulcanised, the joint being fully as strong as the original piece. Retaining-points are not required. For hurried repairs, where a a good dovetail joint can be made, ordinary amalgam will often be found useful and more satisfactory than the fusible metal and cements which have been sold for that purpose.

Wax.—The hard, brown modelling wax for vulcan-ite work takes much better impressions of the mouth than the best English beeswax. When wax has once been in the mouth, it should not under any circum-stances be used again, but thrown in a bow or drawer. When a sufficient quantity is accumulated, melt it in an enamelled iron bowl over water, and allow it to set. When nearly hard, cut the cake out, scrape or cut off the dirt, which will be all found on the under side, and melt again without water. If the heat is too great, bubbles will form on the surface, which can be imme-diately got rid of by sprinkling a few drops of spirits of wine on them. When free from bubbles, pour into thin cakes on ordinary dinner-plates, which have been previously well rubbed with soap and water to pre-vent adhesion. The cakes obtained will be equal, if not superior to new wax. Wax is improved by the addition of a small quantity of dry, brittle Canada balsam. Fresh soft balsam spoils it, and therefore the turpentine it contains must be driven off by heat before it is used.

DR. C. H. LAND'S

COMPOUND

GAS OR GASOLINE FURNACE

(PATENT APPLIED FOR.)

Size No. 1, especially adapted for all kinds of muffle work, crucible work, blow-pipe work, forging and brazing. It is the most complete furnace ever devised for the chemist, assayer, jeweler, dentist and metallurgist. The range of work that can be accomplished with it is practically without limit. Iron, brass and steel castings weighing from 2 to 11 pounds can be made in from 7 to 30 minutes. A muffle 8 inches long, $3\frac{1}{2}$ inches wide, $2\frac{1}{2}$ inches high, inside measurement, can be heated to over 3240° F. in 25 minutes, sufficient to melt wrought iron. Fig. 1 represents the Furnace closed and ready for muffle work. A A is iron pipe, capable of both a sliding and swinging motion (see L Fig. 2) to which the door or plug is securely attached. There is a small hole in the door, covered with a piece of mica, through which all operations can be seen. Observe that the iron pipe is connected to rubber tubing B and with pipe having an air cock C, which regulates the quantity of air passing into the mouth of the muffle. It will also be noticed that the pipe passes over the two holes D D, thus by the escaping flame the pipe is heated to redness and provides a superheated air before reaching the muffle; this column of air forced into the muffle keeps up a counter pressure within, so much greater than the pressure produced by the blast within the fire chamber, that all foul gasses are prevented from entering the muffle even though it is cracked, thus the most delicate porcelain can be baked without the least danger of so-called gasing. Also it will be seen that by connecting the rubber pipe with retorts or gasometers any desired vapor or gas could be forced into the muffle, thus making the Furnace invaluable for scientific experiments.

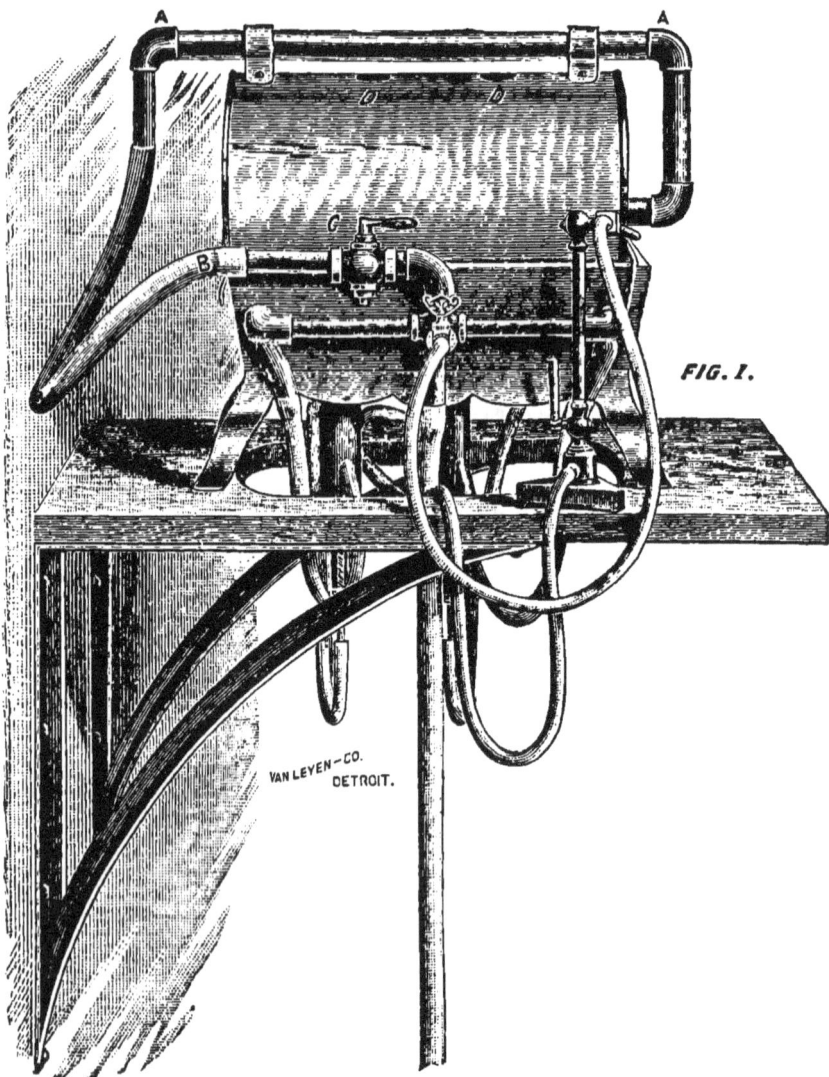

FIG. I.

VAN LEYEN—CO.
DETROIT.

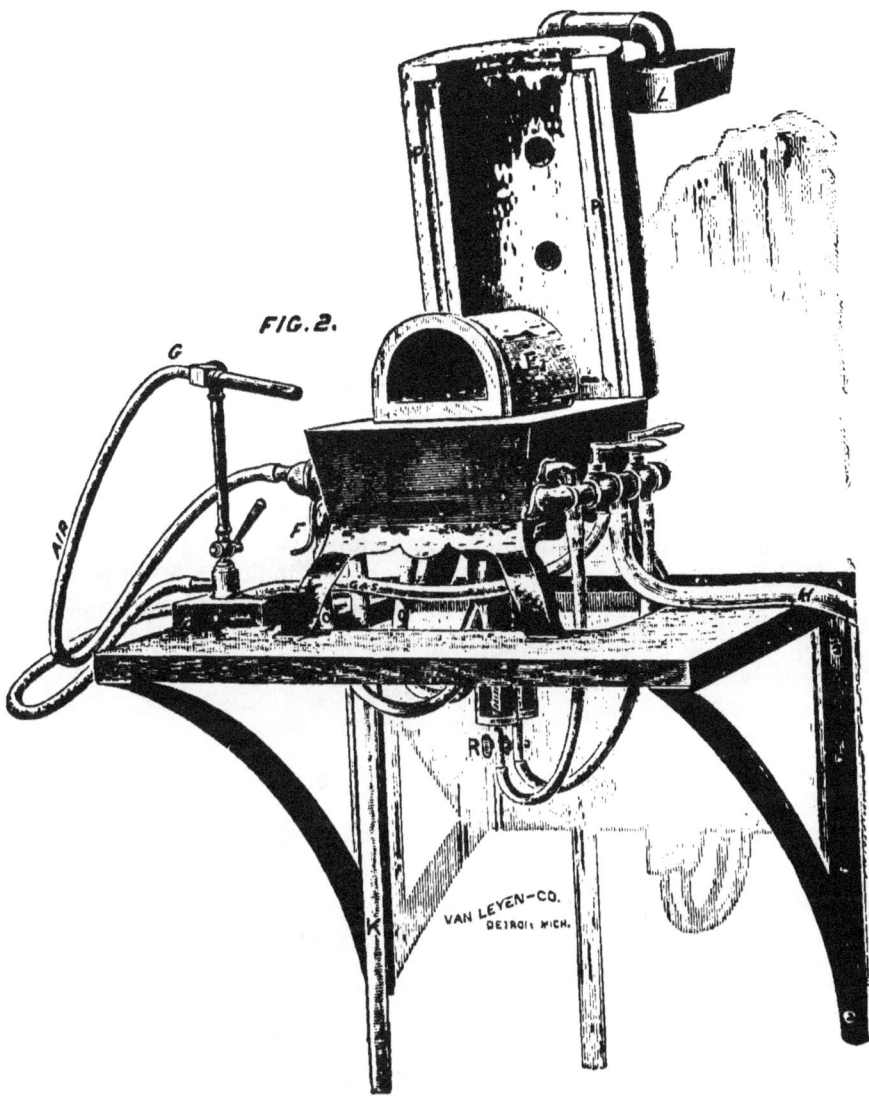

FIG. 2.

VAN LEYEN-CO.
DETROIT MICH.

Fig. 2 illustrates the Furnace thrown open, being swung on hinges at the back, exposing the muffle E. The groove P P is packed with asbestos fibre, so that when the sections are brought together the Furnace will be perfectly air and gas tight. The hooks F F are to hold the upper section secure to the lower. The gas and air connections are so arranged that the ordinary blow-pipe can be attached as shown at G. When the muffle E is removed it exposes two burners and a fire-brick surface made to fit the various appliances for crucible, ladle and blow-pipe work. One or both burners can be operated in conjunction with the blow-pipe G. The air cock R is to provide a means for shutting off the air supply from either burner when required. H is the gas supply; K, air pipe connecting with the bellows. Size of muffle, inside measurement, 8 inches long, $3\frac{1}{4}$ inches wide, $2\frac{1}{4}$ inches high. With gasoline gas porcelain teeth can be enameled in from 10 to 15 minutes; ordinary city gas in from 15 to 25 minutes, according to quality. In 30 minutes a heat sufficient to destroy the muffle can be produced, which indicates a temperature of 3240° F., much higher than is ever needed for all kinds of work, except the fusing of platinum. $\frac{3}{8}$ inch gas pipe will supply sufficient gas and can be worked with ordinary foot bellows.

PRICES.

No. 1 Furnace with Muffle, two Slides, one Crucible Jacket, - - -	$40 00
Bellows, - - - - -	8 00
Blow Pipe, - - - - -	3 00
Extra Muffles, each - - -	1 00
Extra Crucible Caps, - - -	10
Extra Crucible Jackets, - - -	50
Walnut Bracket, - - - -	6 00
Gasoline Generator, - - -	14 00

Any ordinary hole in the bench will answer the same purpose as a bracket.

DETROIT DENTAL MANUFACTURING CO.,

301 Woodward Avenue,

DETROIT, MICH.